BEI GRIN MACHT SICH IHR WISSEN BEZAHLT

- Wir veröffentlichen Ihre Hausarbeit, Bachelor- und Masterarbeit

- Ihr eigenes eBook und Buch - weltweit in allen wichtigen Shops

- Verdienen Sie an jedem Verkauf

Jetzt bei www.GRIN.com hochladen und kostenlos publizieren

Modular Heat Dissipation in a CubeSat. Identifying potential solutions

Anurag Paudel

Bibliografische Information der Deutschen Nationalbibliothek:

Die Deutsche Nationalbibliothek verzeichnet diese Publikation in der Deutschen Nationalbibliografie; detaillierte bibliografische Daten sind im Internet über http://dnb.d-nb.de abrufbar.

ISBN: 9783346764294
Dieses Buch ist auch als E-Book erhältlich.

Modular Heat Dissipation in a CubeSat

Anurag Paudel

Bachelor of Engineering
Mechanical Engineering (Honours)

School of Engineering Macquarie University

June 1, 2022

ACKNOWLEDGMENTS

I would like to acknowledge the support and guidance of Dr. Ediz Cetin throughout this research. His organizational and leadership skills have help me immensely in putting together various ideas and findings into a presentable document.

ABSTRACT

CubeSat has been a favourable means of performing research in space environment for students and universities seeking economically feasible access to space. CubeSats provide great flexibility in regards to utilizing COTS equipment for development. As such, extensive tests need to be performed in order to validate fail-proof operation before launch. Thermal control systems are one such subsystems that need to be taken into consideration. This report aims to assess and analyse the existing methods used in thermal management of CubeSats and identify a potential solution by weighing existing heat-dissipation technologies against each other in their shortcomings and advantages.

Contents

List of Tables

List of Figures

Chapter 1

Introduction

In the late 1990s, Jordi Puig-Suari of California Polytechnic State University and Bob Twiggs of Stanford University were attempting to assist students in gaining technical experience with satellites. They were generally costly to develop and launch and as a result, they introduced the concept of CubeSats. [1]

CubeSats belong to the nanosatellite category of research spacecraft. CubeSats are constructed according to conventional dimensions (Units or "U") of 10 cm x 10 cm x 10 cm. They are commonly 1U, 2U, 3U, or 6U in size and weigh less than 1.33 kg (3 lb) per U. They are launched from a NASA designed deployer pod known as Poly-Picosatellite Orbital Deployer, or P-POD [2].

Russia's Plesetsk launch station launched the first six CubeSats in June 2003. According to [1], the average cost of launching a CubeSat was around $40,000, a compared to millions of dollars to launch a conventional satellite bargain, this was a bargain.

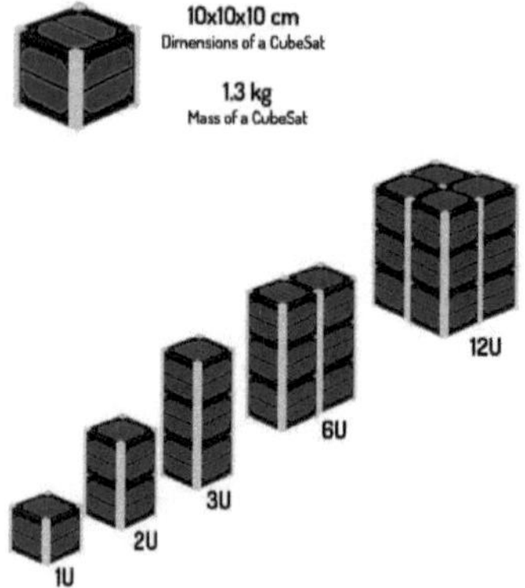

Figure 1.1: CubeSat standard dimenesions [2]

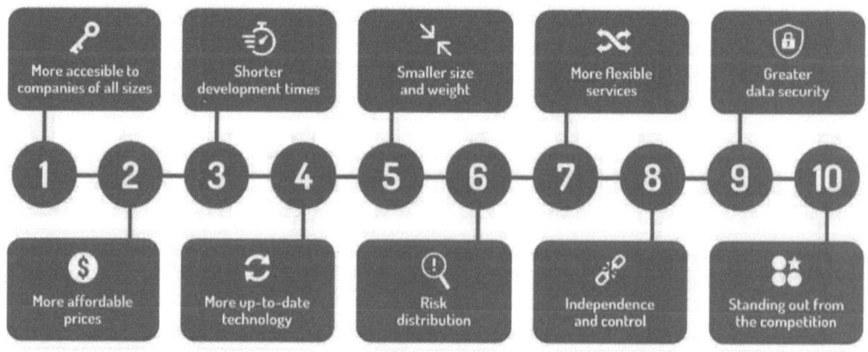

Figure 1.2: CubeSat advantage [3]

1.1 Macquaire's CubeSat Mission

Macquarie is developing its own CubeSat with a projected launch in 2023. In addition to developing a knowledge base at the institution, MQube-1 will conduct tests and demonstrations of important support technologies and comprise of a deployable solar panel array, attitude control electronics, experimental payload using the NVIDIAS Jetson Nano 8GB and fault-tolerant circuits designed for beyond-LEO environments. MQube-1 will be equipped with a specialized thermal infrared imager to conduct mineralogy studies of both the earth and the moon from the SSO, in addition to a number of other secondary instruments. These instruments are designed to capture data that is valuable to research in addition to serving as proof of concept for more complicated future missions. The CubeSat will utilize thermal-IR imager that will be used. Progress is being made on creating techniques and a payload for a CubeSat in order to evaluate unique protocols for hardening microorganisms to withstand the rigorous environment of the space.

1.2 Research Scope

Because of the increased sophistication of the mission risk management, the development of effective thermal management systems has become a key research goal for the MQube-1 designers. This is essential in order to provide them with advice on the efficiency of their design approach. This report aims to assess and analyse the existing methods used in thermal management of CubeSats and identify a potential solution by weighing existing heat-dissipation technologies against each other in their shortcomings and advantages.

1.3 Research Plan

This semester's study on the history and technique of thermal management, solution identification, and feasible solution combinations will develop a knowledge basis. This foundation may be used and expanded upon to further complicated analyses with more precise models and/or mission criteria.

The statement of the Thesis A plan is broken down into two stages, and each phase contains two distinct milestone and objective groups that must be accomplished. In the first part of the project, background study will be conducted on the several cutting-edge thermal management technologies available. In the second phase of the project, which will establish the most efficient thermal management system for MQube-1's primary structure and payload, the results of the research will be used.

Chapter 2

Background and Related Work

2.1 CubeSat Thermal Regime

All components of the spacecraft have a temperature tolerance range within which they must be kept in order to continue meeting the operational and survivability criteria outlined for each phase of the mission [4]. Same applies to CubeSats. The amount of heat that is taken into, stored within, and released from the spacecraft is what determines the temperature of the spacecraft. The flow of heat into and out of an orbiting satellite is shown in an oversimplified form in figure 2.1.

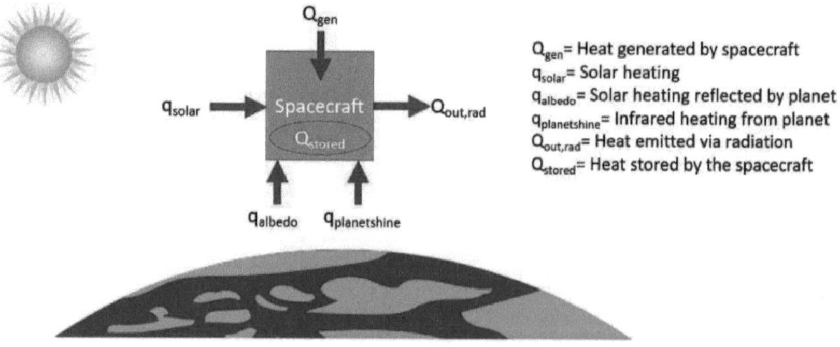

Figure 2.1: Spacecraft thermal regime [4]

2.2 MQube-1 Thermal Profile

Srinivasan [5] presents a detailed analysis on the external heat sources and its effect on the MQube1 CubeSat. Thermal analysis on the primary and secondary (payload) structure has been performed to identify potential issues due to the structural deformation that occur during the worst hot and cold cases as the CubeSat goes around the earth. A total Radiative Heat Flux Regime (RHFR) plot illustrates the sum total of incident heat flux by means of direct, solar and albedo radiation.

Figure 2.2: Total RHFR plot of MQube-1.

The X axis, indicated by the blue arrow represents orbit time and it moves through a 92.5 min orbit. The Y axis, indicated by the green arrow indicates date that moves through 1 year starting at vernal equinox. Finally, Z axis, indicated by the red arrow, represents the amount of incident heat flux received.

By utilizing this plot the author has evaluated extreme temperatures that the satellite experiences throughout its orbit. The maximum temperatures are in the range of 40-50^{0}C whereas, the temperatures go as low as -45^{0}C during coldest situations.

Thermal analysis on the internal heat generation has not yet been conducted on the MQube-1. However, we can obtain an approximation of the extremes of temperatures that a CubeSat of similar dimension and/or orbit might develop during operations and down-times. The major limitation of this approximation is that it is based solely on the heat generated by the most power consuming electronic component, which is the NVIDIA's Jetson Nano (8GB). The Jetson Nano is a compact, powerful computer that enables the concurrent execution of numerous neural networks for applications such as image classification, object recognition, segmentation, and audio processing [6].

NVIDIA has provided a detailed thermal specification of various iterations of the Jetson Nano [7]. The specific thermal profile of the specific Jetson series including the hardware throttling and shutdown temperatures are presented in the table below. The power supply range of 10-30 W satisfies the MQube-1 electronic subsystem design criteria wherein 15 W of maximum power supply is available in the system at any instance.

Thermal zone	Thermal sensor	Cooling action	Jetson Xavier NX series
			10W-30W
CPU-therm	THERMAL_ZONE_PLLX	SW throttling	90.5 °C
		HW throttling	94.5 °C
		SW shutdown	96.0 °C
		HW shutdown	96.5 °C
GPU-therm	THERMAL_ZONE_GPU	SW throttling	91.5 °C
		HW throttling	95.5 °C
		SW shutdown	97.0 °C
		HW shutdown	97.5 °C
AUX-therm	THERMAL_ZONE_AUX	SW throttling	90.0 °C
		HW throttling	94.0 °C
		SW shutdown	95.5 °C
		HW shutdown	96.0 °C
AO-therm	THERMAL_ZONE_AO	HW shutdown	109.0 °C
PMIC-Die	PMIC thermal sensor	HW shutdown	120.0 °C
Tboard-tegra	TMP451 local sensor	HW shutdown	
Tdiode-tegra	TMP451 external sensor	HW shutdown	
thermal-fan-est 5	Weighted average of CPU, GPU and AUX	Fan speed control	46.0 °C

Table 2.1: Jetson series thermal specification

Maximum temperature of 120^0C can be approximated from the table above and an approximation of the lowest possible temperature can be made by assuming idle or switched-off state of all electronics in the coldest region of the orbit which has already been evaluated.

It can be observed that a stable and efficient thermal management system is vital in order to prevent the effect of temperature fluctuations on sensitive components and electronics. Few state-of-the art thermal management systems have been studied and developed specifically for application in CubeSats. These systems have been classified into active and passive systems wherein active systems require continual electric and/or mechanical stimulation to operate whereas passive systems do not [4]. Following discussions will explain in detail the underlying theory, mechanism, history and latest development that have occurred in regards to specific systems.

2.3 Passive Thermal cooling

2.3.1. Multi- Layer Insulation (MLI)

Utilizing materials with certain radiative qualities, such as solar absorptivity and infrared (IR) emissivity, allows for the manipulation of the thermal radiation environment. These features are virtually exclusively surface properties that can be altered simply by coating or covering with specialized insulating materials [4]. Thermal insulation acts as a thermal radiation barrier from incoming solar flux and also to prevent excessive heat dissipation. MLI has historically proven to reflect >90% of incident radiation and has been used in many cryogenic applications including the 1970s moon landings due to its lightness and flexibility [3,8, 9,10].

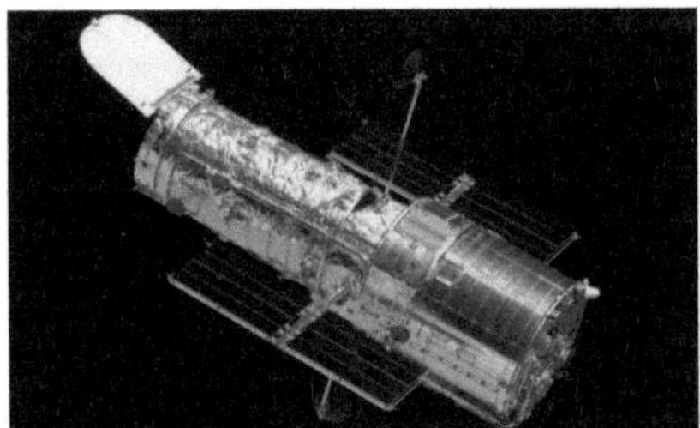

Figure 2.3: Hubble telescope wrapped in MLI [11]

The most adapted form of MLI performance modelling was first derived in a multi-year Lockheed Palo Alto research project that was carried out for NASA in the 1970s [12].

As it can be observed, MLI is fragile and its performance degrades dramatically if two surfaces are not spaced enough creating a thermal short circuit. Hence, it should be used with care or avoided entirely on the outside of miniature satellites that fit in deployers such as P-POD and NLAS. MLI blankets provide a possible entanglement danger in these close-fitting, push-spring-style deployers. In addition, the efficiency of MLI blankets tends to decline as their size lowers because heat transmission through the blanket rises closer to the blanket's edges, and the attachment technique has a significant effect on performance since it generates a heat route. Due of these obstacles, MLI typically performs less effectively on tiny spacecraft like CubeSats. Generally speaking, surface coatings are less sensitive and more suitable for the outside of a tiny spacecraft that will be released from a dispenser [4].

Newly developed Non-Interlayer-Contact Spacer MLI (NICS MLI), employ tiny discrete spacers and may eliminate uncertain interlayer contact between films. Due to the high thermal resistance of this spacer, the NICS MLI blanket exhibits exceptionally good insulation performance with effective emissivity value of 0.0008 between 77K and 300K boundary temperature. At low temperatures specially, the conductive ratio of heat leak through conventional MLI exceeds 89 percent, while the conductive ratio of heat leak through NICS MLIs is less than 8 percent. In contrast, the conductivity of NICS MLI is low, allowing thermal performance to be maintained even at low temperatures. The predicted conductive heat transfer coefficient in each layer for NICS MLI is 3.03E-6 W/K while for traditional MLI it is 7.1E-1 W/K [13].

2.3.2. Paints and coatings

For surfaces that must absorb a high proportion of solar heating and release a large proportion of spaceship heat, matte black paint provides a high solar absorptivity and high IR emissivity. Alternately, matte white paint has low solar absorptivity and strong IR emissivity for surfaces that must take a little amount of solar heating and radiate a large amount of spaceship heat. Silver Fluorinated Ethylene Propylene (FEP) tapes sre highly effective as a second surface absorb incoming solar energy due to low solar absorptivity while releasing spacecraft heat energy effectively due to high IR emissivity. Depending on the application, paints, coatings, and tapes are chosen. Paint is often simpler to apply and remove, less costly, and has a shorter lifespan than tape. However, certain paints, like as Parker-Aeroglaze Lord's 306/307, are costly and need long and highly specialized application methods. Different alternatives may have varying temperature restrictions. All of these considerations must be taken into account in relation to the required application when picking the final solution [4].

In satellites, atomic oxygen (AO) and ultraviolet (UV) radiation have a negative impact on paints and other thermal coatings. The significance of these factors increases when the synergistic effects of AO and UV are included. In a case study of a satellite with A276 white paint, which has an uncommon 100 percent degradation rate and initial absorptance value, the satellite radiator suffered a maximum temperature rise of around 12 °C near the end of its mission. This significant rise in satellite radiator temperature recommends that, during the design phase, a suitable margin for the absorptance value at the beginning of life must be considered to allow for the deterioration of this paint at the conclusion of the satellite mission. This is also undesirable from the standpoint of satellite power consumption since it raises the satellite power budget to compensate for the higher power consumption in the heaters [14].

2.3.3. Thermal straps

Thermal straps are flexible, thermally conductive connection used to transport heat. They're employed between heat-generating chips and a chassis wall or radiator. Their flexibility prevents additional structural loads. Thermal straps may be composed of copper, aluminium, or graphite. Large spacecraft employ straps with more than two end blocks and various types materials [4]. Heat conduction between solids is calculated in terms of thermal contact conductance (TCC). Pressure and surface flatness have a significant impact on this characteristic. TCC has an impact on the thermal strap's overall performance [9, 15].

For thermal subsystems, graphite fibre thermal straps (GFTS) provide the optimal mix of attributes, including high heat conductivity, low mass, and low stiffness. For systems requiring operational temperatures below 40^C such as cryocoolers, conventional metal straps such as aluminum and copper are not feasible within the mass or volume constraints imposed. GFTS have more than 10 times the thermal conductance of copper straps and more than 7 times the thermal conductance of aluminium straps of the same weight while working in the 300 K temperature range [18]. Based on thermally tested and produced thermal straps, Figure 2.5 best illustrates the superior thermal conductance of graphite thermal straps against metal straps on a weight-for-weight basis.

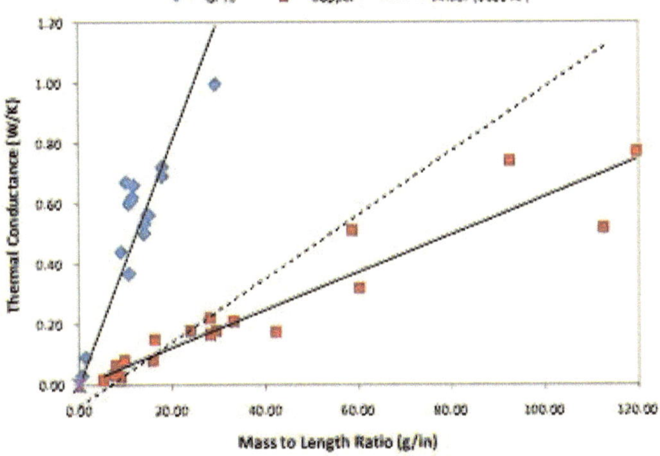

Fig 2.5: Thermal conductance of graphite thermal straps against metal straps on a weight-for-weight basis [18]

Space Dynamics Laboratory (SDL) has supplied Utah State University with a Pyrolytic Graphite (PGS) strap for the Active Thermal Architecture (ATA) project [3,19]. Temperature conductance testing of 0.001" thick aluminium, 0.001" thick copper, and 0.001" thick PGS was conducted to compare and contrast thermal performance across a wide temperature range. Thermal conductivity of copper and aluminium foils surge at cryogenic temperatures, whereas the PGS functions best at temperatures over 70K. Cryogenic aluminium straps function substantially better than copper since their density is 30 percent lower and the PGS performs even better at higher temperatures with a density 50% lower than aluminium. Specific performance will vary according on strap shape; nevertheless, these temperature-dependent patterns apply to all straps made with these materials [16].

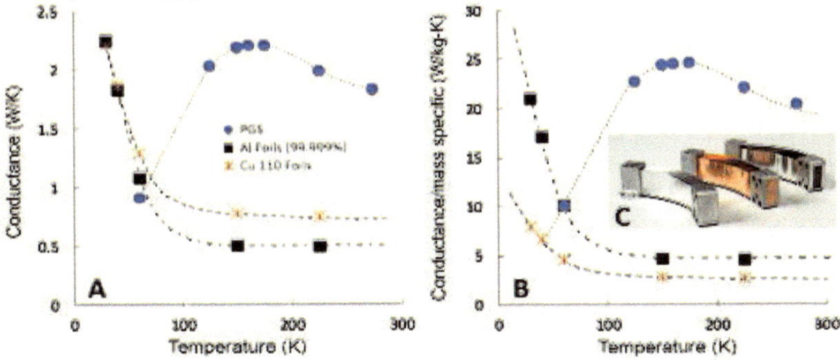

Figure 2.4: Left- thermal conductance and Right- mass specific thermal conductance between (PGS, Copper and Aluminium (99.99% purity) [16]

i2C Solutions is developing FlexCool, a lightweight two-phase heat strap. Using acetone as the working fluid, effective thermal conductivities of up to 2,150 W/m-K were shown using a 0.86 mm thick heat strap. The heat strap can endure internal vapour pressures as high as 930 kPa, demonstrating its capacity to operate in space. Thermal strap is 400% more conductive than a pure copper spreader [17].

Pyrovo™ has developed thermal straps that claim to have have over 4x the thermal conductivity of equal sized copper thermal straps and 7x that of aluminium. This implies that the same conductance may be attained with a fraction of the volume, mass, and rigidity. This is particularly crucial for the cooling of optics and fragile electronics. Moreover, thermal conductivity is much greater at cryogenic temperatures [20].

MATERIAL COMPARISON

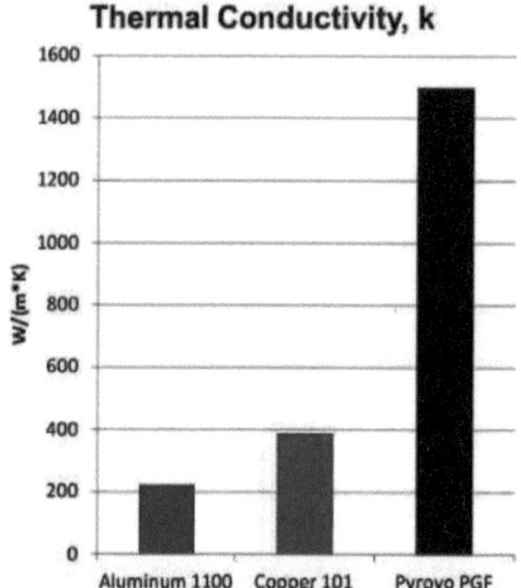

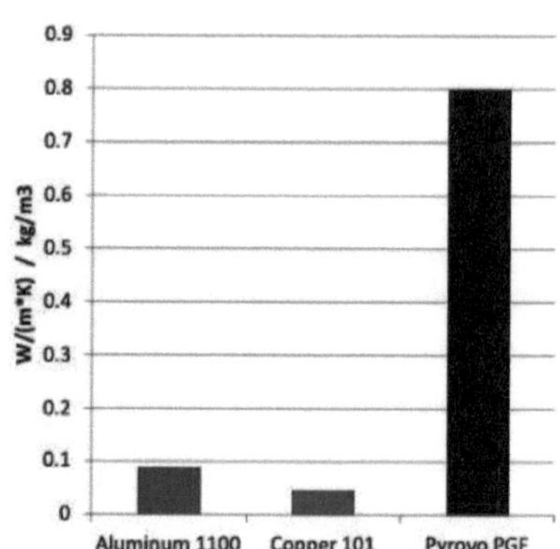

Figure 2.6: Pyrovo PGF compared against common materials

2.3.4. Thermal interface materials

In order to improve the conductive heat transfer between two components, terminal interface materials are put between them. Thermal conductivity, temperature restrictions, and vacuum-compatibility all vary among the many kinds of materials used to make a sheet or pad for sandwiching between two surfaces. Between heat dissipating electronics boxes and mounting surfaces, thin sheets of materials are widely utilized to cool the electronics by transferring the heat to a cooler surface. A particular amount of pressure must be maintained between components in order for these materials to function properly [4].

TflexTM HD80000 from Laird has a thermal conductivity of 6 W/mK and a higher pressure-to-deflection ratio. In addition to reducing component stress, this combination also reduces heat resistance. In spite of its softness, HD80000's high thermal performance may be maintained by handling and applying the material by hand without the addition of fibreglass or another support layer [21].

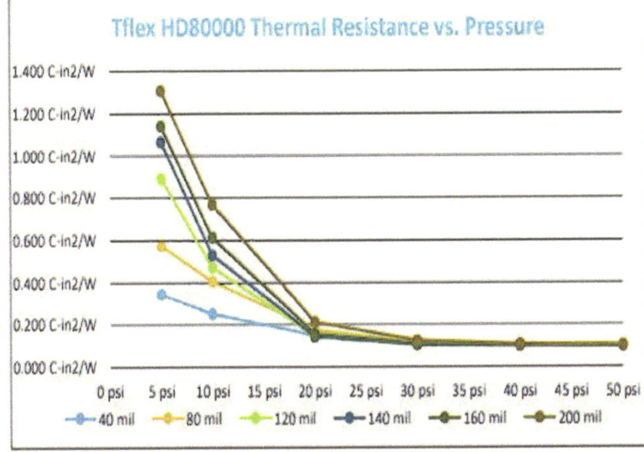

Figure 2.7: TflexHD80000 Thermal Resistance vs. Pressure [21]   Figure 2.8: Thermal interfacing material (Tflex) [4]

2.3.5. Passive and deployable radiators

In space, waste heat from spacecraft components can only be be lost by radiative heat transfer. Radiator panels are developed for this function and are available in single active face (BMR) and deployable variants (DR) [22, 23].

By knowing internal heat generation and other ambient heat loads, radiator area may be determined using Equation [25, 26]. Albedo flux and IR flux are often omitted in satellite radiator design estimates since the earth's albedo and IR flux are minimal (roughly 0) as altitude rises [24].

Based on this equation, it is possible to deduce that the total heat rejected by the deployable radiator is more than that rejected by the BMR, which only has a single-active radiating surface. This is because the deployable radiator has double-active radiating surfaces. For the sake of the calculations shown above, it has been assumed that the radiating surfaces are painted black, which has an emissivity of 0.85 [27].

Results indicate that deployable radiators significantly improve the amount of heat a satellite can disperse, particularly. Using deployable radiators may result in a practical dissipation of up to 200 W for a notional 6U CubeSat, as opposed to just 90 W for body-mounted radiators. This corresponds to a possible doubling of the permitted bus power for CubeSats. Using deployable radiators will enhance the overall bus power since more heat can be successfully dissipated, validating deployable radiators as an essential component of the future of high-powered CubeSats. Therefore, it is of great relevance to pursue development of this technology to increase the radiator's surface area [27].

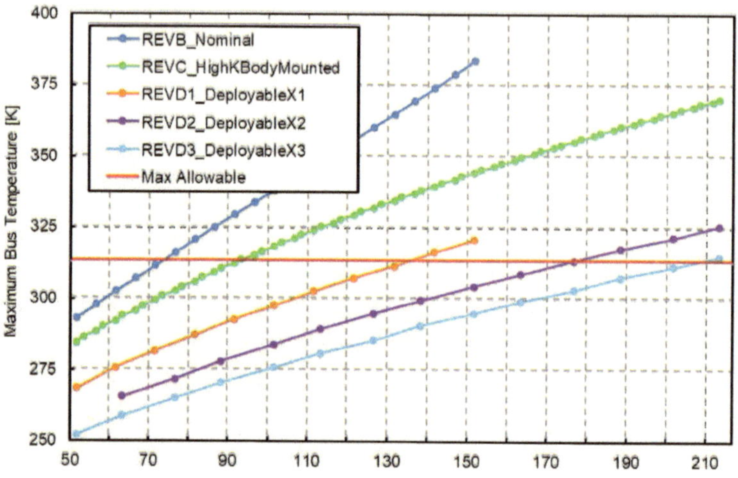

Figure 2.9: Maximum bus temperature vs wattage with different radiator types [27].

Shoya Ono, Hosei Nagano, and colleagues from Kaneka Corporation and JAXA created a flexible spacecraft radiator in 2015. Depending on external conditions, this design deploys or stows the radiation area [28] in 0.5 x 360 x 560 mm form factor and weighs 0.287 kg. A shape memory alloy and bias spring actuator stores and deploys the radiator. Kaneka Graphite Sheets (KGS) are used to improve radiator size and heat conductivity. The fin's back is insulated with MLI to decrease heat loss in cold weather. The half-scaled radiator dissipated 54 W at 60°C in deployment and stowage tests in a thermostatic chamber [4].

2.3.6. Passive Heat Pipes

For space applications, heat pipes are one of the most effective passive thermal management systems since they do not need any extra electricity to operate. These devices have tiny cross sections and excellent thermal conductivity, allowing them to transport heat across great distances inside the satellite. Evaporator, adiabatic, and condenser are the three main portions of a heat pipe. The evaporator segment of a conventional heat pipe receives heat from a heat source. Temperature rises in the working fluid's temperature range and causes it to change phase from liquid-to-vapor, moving it to the adiabatic or transportation section, where it is condensed. The latent heat of this fluid is subsequently rejected to the sink. [4].

As a result of capillary pumping action, the condensed liquid travels back to the evaporator part. The heat load on the heat pipe determines the size of the pipe [28, 29]. Thermal conductivity may be adjusted by modifying the physical parameters such as pore diameter, porosity, and permeability of Constant Conductance Heat Pipes (CCHP).

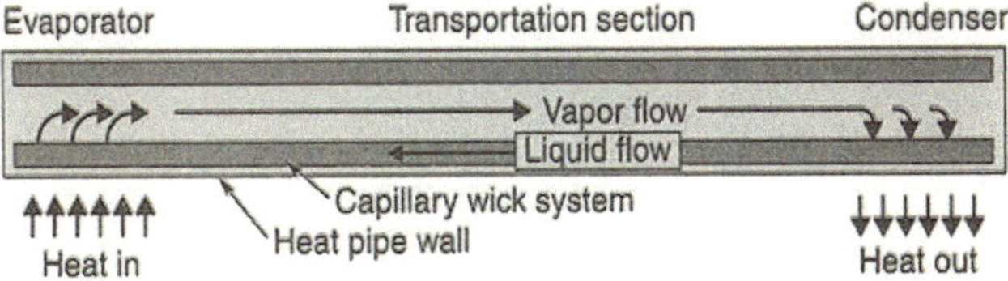

Figure 2.10: Common heat pipe schematic [4].

The fluid selection in heat pipes is based on equations below that define their capillary limit [30]. M is the figure of merit quantity that provides a figurative comparison between different choices of working fluid in different ranges. Water is the most efficient working fluid that is based on this merit.

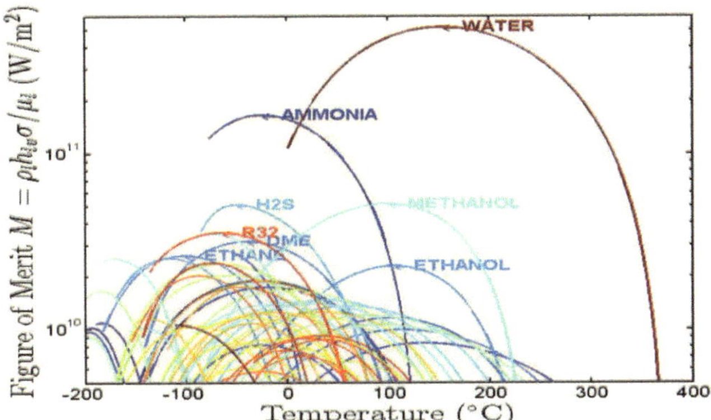

Figure 2.11: Heat pipe working fluid figure of merit [30].

2.3.7. Phase change Material- Heat Storage Device (PCM-HSD)

PCM thermal control converts thermal energy into a phase change reaction, storing heat while it's created and releasing it when the electronics is off. Since the phase change process happens at virtually constant temperature, thermal management means the system temperature doesn't fluctuate considerably during melting/solidification, therefore the electronics may be effectively protected. [31].

In order to select appropriate PCM material, set of equations may be used depending on the physical (dimensional or mass) limitation that is imposed. An example with a weight limitation of 0.5 kilogrammes to determine the minimal latent heat required for the PCM is found to be 81kJ. This value however is not accurate and the PCM's latent heat has to be increased since it will be housed in a container that adds mass yet has essentially little latent heat [31].

Once the PCM-HSD's overall dimensions are understood, latent heat energy may be calculated. If this energy is more than 81kJ (162kJ/kg0.5kg), the chosen PCM can be considered suitable [31]. Two materials, RTSHC and $KF4H_2O$ were found to be the most efficient materials in terms of usability using PCM-HSDs. It can be observed that even at the critical point of design, the PCM-HSDs did not let allowable temperatures exceed the maximum allowed temperature of 40^0C, for this particular arrangement.

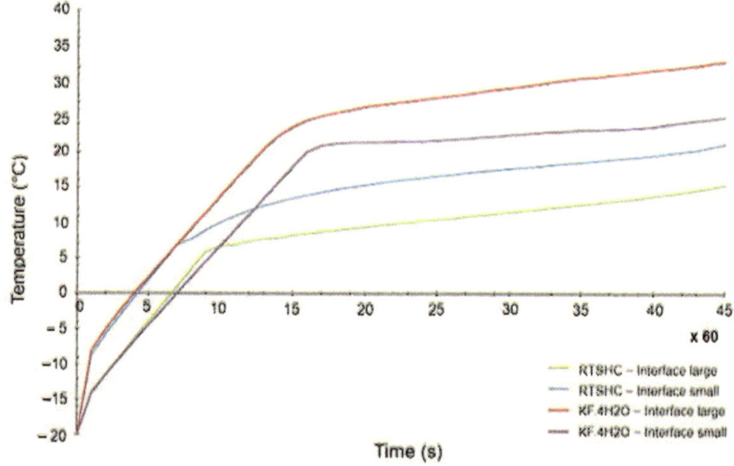

Figure 2.12: Peak temperature plot for RTSHC and KF4H₂O during transient operation [31].

2.4 Active Systems

2.4.1. Cryocoolers

Cryocoolers are used to cool instruments or subsystems such IR sensors usually below 100K [33]. Imaging spectrometers, interferometers, and midwave infrared (MWIR) sensors need cryocoolers as low temperature enhances wavelength coverage and dynamic range. Cryocoolers provide extended instrument lifetimes, minimal vibration, great thermodynamic efficiency, low mass, and cooling temperatures below 50K [32].

Creare has developed an Ultra-Low Power (ULP) single-stage turbo-Brayton cryocooler that runs between cryogenic and load temperatures. This cryocooler can run at 30 to 70K with loads up to 3 W and 210K by altering charge pressure and turbo machine speeds [4]. The K562S is a rotary Sterling small micro-cooler designed by Ricor-USA, Inc. Cooling capacity ranges from 200 mW at 95 K to 300 mW at 110 K. Several tiny military gimbals have used this technology.[4]. Sunpower [35], Northrop Grumman [36], Thales Cryogenics [37], have also developed state-of-the-art cryocooler suitable for CubeSat usage.

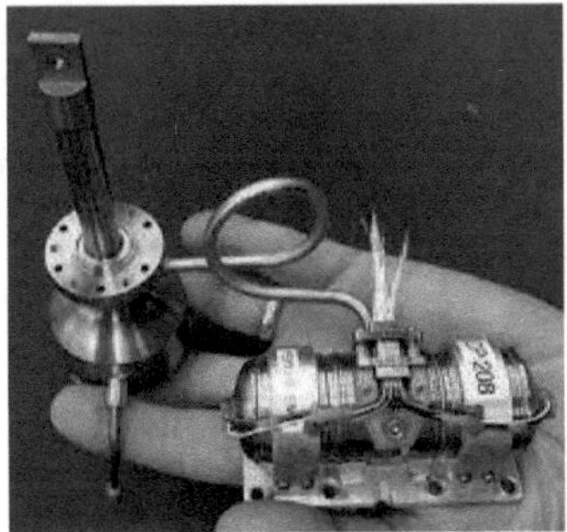

Figure 2.13: Miniature cryocooler developed at Lockheed Martin [34].

However, special attention must be made to the cryocooler's electrical use in order to minimize any cooling concerns that arise due to a lack of electrical power. In order for the cryocooler to work correctly, it must be coupled to a radiator that can effectively remove the heat it generates [33]. The size of the radiator will be dictated by the size of the cryocooler, as their rejected power varies [4].

2.4.2. Thermoelectric pelteir cooling

When an electrical load is applied to two distinct temperature points on a conductor, current flows. Depending on materials and current direction, one material junction absorbs heat while the other emits it. One point is cooled and the other is heated. This is called the Peltier effect and is employed in cooling and temperature control equipment [3,33].

There are three main requirements on thermoelectric materials concerning an efficient power conversion [38]: High Seebeck coefficient, low electrical resistance and low thermal conductivity. Currently, semiconductors are the most preferred class of materials for peltier applications because they fit the criteria, but within this class there are variances that impact optimal performance temperature and as such there is no all-in-one solution for a large temperature range. Bismuth telluride (BiTe) is best for satellite heat fluxes since temperatures are below 100^0 C [39].

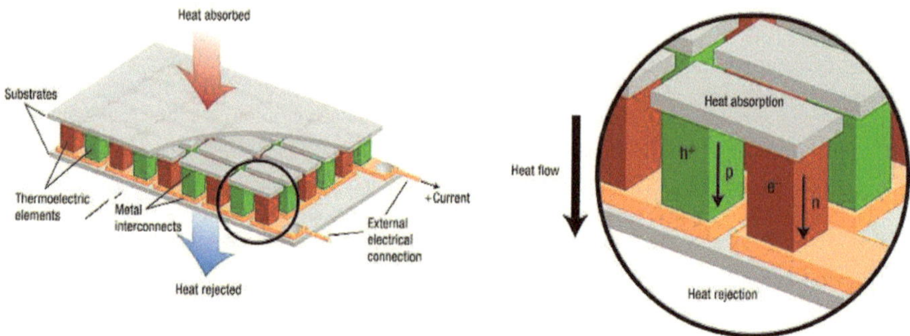

Figure 2.14: Schematic of thermoelectric peltier generator [40].

Even though several integration points have been examined, CubeSat's energy harvesting capability is limited by current technology. In all situations, the integrated TEG's temperature gradient is less than 3K. To maximize thermal potential, two facing solar cells should be directly linked by a TEG. This is simply a theoretical technique because of the amount of work and power required to implement it. Using heat in tiny satellites is difficult because of their small size. As a result, the heat transfer is quite diffuse [39].

2.4.3. Micro heat sink and pump

Micro-technologies are needed to prevent thermal spots, develop more efficient and compact heat pipes, and disperse heat locally. CubeSat's bulk and power limits favour passive thermal control technologies, notably heat pipes. The space industry's 40-year use of such instruments makes their development for micro- and nano-satellites inevitable. Heat pipes are embedded components having a thermal conductivity tens of times greater than aluminium or beryllium and as such can be used to eliminate hot spots and transfer thermal power evenly throughout a satellite's construction [42].

Two types of silicon micro-heat pipes were examined by Le Berre et al. [41]. Their triangular channels, each with an area of 0.02 and 0.09 mm^2, were created using an anisotropic method. Methanol was used to power the MHP. With a 600 W/mK effective conductivity, the micro-components performed quite well. Kang et al. produced an unusual and intriguing material selection [43]. For the gas and liquid parts, they used three-layer copper pipes. Diffusion-based bonding was used to join the three layers together. The greatest results were obtained using an MHP that was 80 percent methanol loaded [44].

The need for a pump is one of the primary issues with this method. This necessitates a very compact pump system with unproven dependability for the CubeSats. Vibrations, are major threats to CubeSats and this constitutes a further issue. Despite this, several studies attempt to combine mechanical fluid systems. In addition, the thermal equipment must be resistant to the harsh launch environment. Therefore, this thermal control system's design is a compromise between thermal and structural criteria [44].

2.4.4. Mini-mechanically pumped fluid loops

In a mechanically pumped fluid loop (MPFL), a heat exchanger and a heat sink are coupled to a circulating pump, which circulates a liquid via tubing. The heat exchanger is used to transfer heat from a heat source to a heat sink, usually a radiator, and then to return the cooled fluid to the heat source to continue cooling it further. Forced convective cooling may be used to cool numerous places using an MPFL. Because of their large bulk and significant power consumption, they are seldom utilized on CubeSats.[4]. However, lots of research have been performed to design and develop such fluid-loop systems for micro form-factor spacecrafts.

A mini-MPFL for a closed-cycle Joule Thomson cryocooler is being developed by Lockheed Martin Corporation [4]. To circulate gas, it requires just 1.2 W of electrical power, and it can handle roughly 40 W of spaceship power as a single-phase loop, or several hundred Watts of spacecraft power as a two-phase loop. With a total mass of 0.2 kg, it is able to do both. In a 2016 research, the compressor was tested and found to have a compression efficiency of 20–30% [45].

A multi-parallel micro pump is the core of the mini-MPFL with 10-30 pumps running simultaneously, a single-point failure may be avoided by using this design. Even if one pump fails, the other n-1 pumps continue to provide flow, while the flow declines to only a fraction of (n-1)/n of the original flow [46].

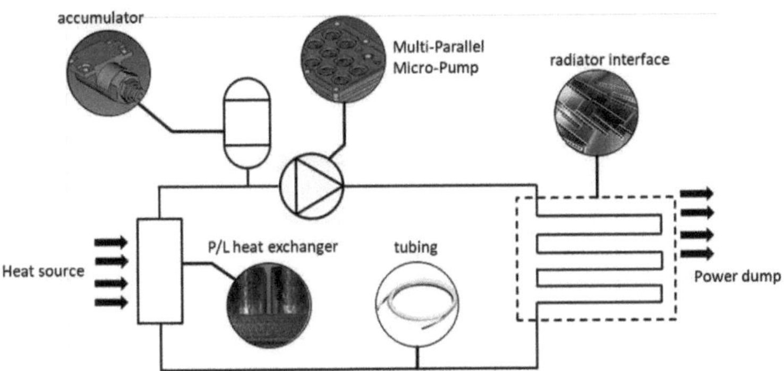

Figure 2.15: Schemcatic of the multi-parallel-micro-pump, MPFL [46].

Working fluid selection for this system is similar to the selection process in any heat pipes and is based on a figure of merit. The most efficient fluid for this purpose has highlighted in green in the table below with yellow representing the next best choice.

	$M_{\Delta p}$	M_{Pump}	M_{acc}	Weighted Result
Weights	1	2	4	
Fluid name				
R134a	1	1	1	1
R142B	0.78	0.65	1.64	1.73
NOVEC 7000	0.77	0.78	1.99	2.06
R141B	0.62	0.51	2.61	2.42
NOVEC 7500	0.51	0.53	3.32	2.97
Galden HT80	0.39	0.33	3.65	3.13
Galden HT110	0.42	0.4	3.7	3.20
Opteon SF10	0.39	0.35	4.36	3.71

Table 2.2: Figure of merit for working fluids [46].

Chapter 3

Identifying potential solutions

3.1. Constraints

Selection criteria for any of the systems that have been discussed will be bound to the mass and dimensional constraints imposed by CubeSats. These constrains and challenges have been perfectly identified by NASA [4].

SmallSat Property	Challenge
Low thermal mass	The spacecraft is more reactive to changing thermal environments.
Limited external surface area	There is less real estate to be allocated to solar cells, designated radiator area, and/or viewports required for science instruments.
Limited volume	There is less space for electronic components, science instruments, and thermal control hardware. Components can be more thermally coupled.
Limited power	There is less power available for powered thermal control technology.

Table 3.1: CubeSat properties and challenges imposed [4].

3.2. Assumption

The main objective of this research is to derive a potential, most efficient method to dissipate heat generated inside the MQube1 CubeSat by its electronic components and also shield the satellite from external radiation. Establishing a precise thermal profile and evaluating accurate extremes of temperatures is out of scope of this research as it demands advanced computational modelling and FEA. Computational analyses will explicitly require a much sophisticated numerical models to begin with, since both radiative and conductive heating effects in conjunction need to be taken into consideration and up until now no attempt has been made in deriving a well-established thermodynamic relationship between these two heat transfer modes on a CubeSat and surrounding system.

As such, potential solutions will be based around their effectiveness in preventing the effect of either internal or external source individually. This essentially results in most passive systems being considered for barricading external flux and only a select few being considered to dissipate heat generated by the payload. It is important to note that none of the active systems mentioned in the background research can be used to shield the satellite from external radiation and hence their evaluation will only be based on the internal thermal profile.

3.3. Methodology

Using above thermal analysis data, we can begin deducing a potential solution using comparison tables and weighed measurements. The following discussions will highlight the main features of each systems mentioned in the previous chapter. Next, these systems will be tabulated to compile a numerical rank based on theoretical efficiency, reliability and economic viability. The rationale behind these properties will all be direct references to the background research conducted in chapter 2. The tables will be then gradually narrowed down to a select- few, most promising systems. Then, the best system will be discussed in detail, additionally, in terms of integration with the MQube-1 project. Finally, a possible combination of systems that could facilitate any added future requirements will also be discussed.

3.4. Analysis, evaluation and discussion

3.4.1. System capability, advantages and disadvantages.

System ID	Thermal characteristic	Advantages	Disadvantages
2.3.1	Latest development in MLI technologies have obtained conductive heat transfer coefficient in each layer upto 3.03×10- 6 W/K. >90% efficiency in preventing external radiation.	Extremely lightweight, Ease of installation and integration, Historically proven efficiency.	Performance drops drastically if compressed and with decrease in size, Snagging hazard in tight-fitting deployers.
2.3.2	Paint and coatings properties are material dependent. Some tapes, can be used with MLI to increase effective insulation.	Low production cost, Ease of application on satellite surfaces.	AO and UV affect performance drastically depending on orbital time.
2.3.3	An average of 2,150 W/m-K, for heat loads between 4.5 and 15 watts. State-of-the-art graphite are 7x better than aluminium straps and 10x better than pure copper straps.	Their flexibility prevents the addition of structural loads, Cheap, lightweight, easy integration and highly malleable.	Cannot be used to prevent radiative heat transfer, Graphite can short-circuit electronics if strap is damaged during operations.

2.3.4	0.05 W/MK conduction resistance at 50 psi.	Cheap, lightweight, Compact, Extremely efficient in smaller satellites.	Cannot be used to prevent radiative heat transfer, Efficiency decreases with increase in scale, Not useful unless heat generating components that are fixed on the satellite faces.
2.3.5	Best radiators are able to dissipate 54 W at 60°C.	Can be used in combination with multiple passive and active methods to form an autonomous dissipation system, Can double as solar panels to absorb incident heat.	Cannot prevent radiative heat transfer, Might impose weight constraints on CubeSats.
2.3.6	A 4mm diameter pipe with 0.2mm groove width, can effectively transport a heat load of around 10W at temperatures of -35^0 C to 40^0 C.	Low-complexity, Cheap, well-established, Easy to integrate, Versatility in terms of working fluid, Basis of many active cooling and integration possible with many passive and active systems. (Eg. Heat pipe to radiator, PCM heat pipe system etc.).	Compromised flexibility in smaller spaces, Cannot be used to prevent radiative heat transfer, Cannot provide temperature stabilization during colder orbits.
2.3.7	Best PCM material selected based on figure of merit can store 208J/g of heat energy and at 40 W of dissipated power temperatures can be maintained a t 50^0C which is ideal.	Can store heat, and can be released during cold orbits, No power required.	Cannot be used to prevent radiative heat transfer, Bulky, Require considerable amount of material, not suitable for mass constrained satellites, To cool a 1U area >1.2U

			area of PCM is required.
2.4.1	Can maintain temperatures at 30K to 70K, with loads of up to 3 W, and heat rejection temperatures of up to 210K.	Suitable for high precision imaging devices that require constant low optimal temperatures, Helps expand life of instruments.	Cannot be used to prevent radiative heat transfer, Has a lot of moving parts, in turn increases complexity of the overall system design., These moving parts can induce vibrations into the payload, Requires more power to run efficiently, occupies more size, It is obvious that it produces lot of heat as a result of first law of thermodynamics, Expensive.
2.4.2	No thermal data has been recorded yet in CubeSats.	No moving parts, best for vibration sensitive payloads. Able to cool a component to below ambient temperatures, Very low maintenance.	Cannot be used to prevent radiative heat transfer, Efficiency is low. The flowing current itself tends to generate a significant amount of heat, which gets added to the overall heat dissipation, Requires constant current source to actively cool or heat the components.
2.4.3	Thermal conductivity upto 600 W/mK.	Useful for precise highly localized heat extraction and dissipation.	Cannot be used to prevent radiative heat transfer, Pump system relaibility is still to prove, Induced vibrations threaten operations of CubeSats.
2.4.4	Can handle roughly 40W of spaceship power as a single-phase loop, or several hundred watts of spacecraft power as a two-phase loop. With a	Prevents single-point of failure of a pumped loop. If one pump fails, still n-1 pumps are left to provide flow, which drops relatively to (n-1)/n	Cannot be used to prevent radiative heat transfer.

	total mass of 0.2 kg, it is able to do both.	fraction of the original flow. Compatible with various working fluid and offers flexibility like heat pipes.	

Table 3.2: System capability, advantages and disadvantages.

With these qualities identified and distinguishing grounds laid out, we can assign a numerical ranking to these systems. By analyzing table 3.2, it can be said with utmost certainty that no active system is able to prevent the effects of external radiative heat and analogically no passive system is able to dissipate the heat generated internally. In regards to this, it serves only logical to rank these systems separately based on their type (active or passive).Through such process we shall be able to succinctly allocate the best solutions to the most suitable domain, namely the external and internal thermal management system.

3.4.2. Ranking and rationale

System	System type	Rank
MLI	Passive	1
Paint and Coatings	Passive	7
Thermal Straps	Passive	2
Thermal interface materials	Passive	3
Passive and deployable radiators	Passive	5
Passive heat pipes	Passive	4
PCM-HSD	Passive	6

Table 3.3: Ranking passive systems

Cryocoolers	Active	2
Thermoelectric peltier cooling	Active	4
Micro heat sink and pump	Active	3
Mini- MPFL	Active	1

Table 3.4: Ranking active systems

3.4. Discussion

Through extensive research we have come to a sturdy conclusion with succinct rationale. MLI has been chosen as the best passive system and mini- MPFL has been chosen as the best active heat dissipation system. [46] also provides a comparison table that further helps to reinforce this rationale. and as such this combination of thermal management system holds a lot of potential in regards to being adapted in the MQube-1 CubeSat.

Thermal solution	Comparison with mini-MPL	
	Advantages	Drawbacks
HP's	Simple, low cost, well-known, No active components	Inflexible and non-modular design Limited amount of hot spots can be addressed Rigid connection between PCB connection and CubeSat frame Low reliability of heat switch function
Mini-LHP's	No active components, well-known	Inflexible and non-modular design Limited amount of hot spots can be addressed No or heavy heat switch capability
PCM	No active components, low-cost, well-known,	Inflexible for late orbital changes Limited amount of hot spots can be addressed
Thermal straps	No active components, reliable, low cost	Low thermal performance

Table 3.5: Mini-MPFL vs other state-of the-art active systems [46].

Chapter 4

Conclusion and Future Work

Through extensive research we have come to a sturdy conclusion with succinct rationale. However, this solution is only based on assumption and analysis of existing technologies with custom and non-reviewed decision making processes. Future work will consist of validating the feasibility of this potential solution through the usage of computer modelling and mathematical analysis.

Appendix A

List of Abbreviations

AO	Atomic Oxygen
ATA	Active Thermal Architecture
BMR	Body Mounted Radiator
CCHP	Constant Conductance heat Pipes
COTS	Commercial Off-The Shelf
DR	Deployable Radiators
FEA	Finite Element Analysis
FEP	Fluorinated Ethylene Propylene
GFTS	Graphite fiber thermal straps
HSD	Heat Storage Deviece
IR	Infrared Radiation
KGS	Kaneka Graphite Sheets
MHP	Micro Heat Pipes
MLI	Multi-Layer Insulation
MPFL	Mechanically pumped fluid loops
MWIR	Midwave Infrared
NCIS	Non-Interlayer-Contact Spacer
NLAS	Nanosatellite Launch Adapter System
PCM	Phase Change Material
PGS	Pyrolitic Graphite Strap
POD	Poly-Picosatellite Orbital Deploye
SDL	Space Dynamics Laboratory
SSO	Sun-Synchronous Orbit
TCC	Thermal contact conductance
TEG	Thermo-Electric Generator
ULP	Ultra-Low Power
UV	Ultra Violet

References

[1] "CubeSats: Tiny Payloads, Huge Benefits for Space Research", Space.com, 2022. [Online]. Available: https://www.space.com/34324-cubesats.html. [Accessed: 03- Jun- 2022].

[2] "Basic Guide to Nanosatellites", Google.com.au, 2022. [Online]. Available: https://falenspace/basic-guidenanosatellites/[Accessed: 03- Jun- 2022].

[3] A. Space, "10 Advantages of CubeSats vs. Conventional Satellites", Info.alen.space, 2022. [Online]. Available: https://info.alen.space/advantages-of-cubesats-vs-conventional-satellites. [Accessed: 03- Jun- 2022].

[4] "7.0 Thermal Control", NASA, 2022. [Online]. Available: https://www.nasa.gov/smallsat-institute/sst-soa/thermal-control. [Accessed: 03- Jun- 2022].

[5] N. Srinivasan, Structural Verification of the MQube1 3U CubeSat. 2020.

[6] "New Jetson Nano 2GB Developer Kit", NVIDIA, 2022. [Online]. Available: https://www.nvidia.com/en-au/autonomous-machines/embedded-systems/jetson-nano/education-project [Accessed: 03- Jun- 2022].

[7] Platform Power and Performance — Jetson Linux
Developer Guide 34.1 documentation", Docs.nvidia.com, 2022. [Online]. Available: https://docs.nvidia.com/jetson/archives/r34.1/DeveloperGuide/text/SD/PlatformPowerAndPerformance.html. [Accessed: 03- Jun- 2022].

[8] D. Gilmore, Spacecraft thermal control handbook. Norwich, NY: Knovel, 2007.

[9] S. Selvadurai, A. Chandran, D. Valentini and B. Lamprecht, "Passive Thermal Control Design Methods, Analysis, Comparison, and Evaluation for Micro and Nanosatellites Carrying Infrared Imager", Applied Sciences, vol. 12, no. 6, p. 2858, 2022. Available: 10.3390/app12062858.

[10] Thermal Management of High Power Applications in CubeSat. [Online]. Available: http://dx.doi.org/10.13140/RG.2.2.15695.94881. [Accessed: 03- Jun- 2022].

[11] "About the Hubble Space Telescope", NASA, 2022. [Online]. Available: https://www.nasa.gov/mission_pages/hubble/story/index.html. [Accessed: 03- Jun- 2022].

[12] R. Ross, "Quantifying MLI Thermal Conduction in Cryogenic Applications from Experimental Data", 2022. .

[13] T. Miyakita, R. Hatakenaka, H. Sugita, M. Saitoh and T. Hirai, "Development of a new multi-layer insulation blanket with non-interlayer-contact spacer for space cryogenic mission", Cryogenics, vol. 64, 2014. Available: 10.1016/j.cryogenics.2014.04.008.

[14]   A. Anvari, F. Farhani, K.S. Niaki, Comparative Study on Space Qualified Paints Used for Thermal Control of a Small Satellite. 2009.

[15]   S. Selvadurai, A. Chandran, D. Valentini and B. Lamprecht, "Passive Thermal Control Design Methods, Analysis, Comparison, and Evaluation for Micro and Nanosatellites Carrying Infrared Imager", Applied Sciences, vol. 12, no. 6, p. 2858, 2022. Available: 10.3390/app12062858.

[16]   M. Ralphs, M. Sinfield, M. Felt, High Performance Thermal Straps for a Full Range of Application Temperatures. 2022.

[17]   S.A. Isaacs, D. Arias, M. Hulse, Development of a Two-Phase Heat Strap for CubeSat Applications. 2022.

[18]   S.J. Nieczkoski, E.A. Myers, Highly-Conductive Graphite Thermal Straps Used in Conjunction with Vibration Isolation Mounts for Cryocoolers. 2022.

[19]   "SDL - Capabilities", Sdl.usu.edu, 2022. [Online]. Available: https://www.sdl.usu.edu/capabilities/science-engineering/thermal-management/thermal-straps. [Accessed: 03- Jun- 2022].

[20]   "Thermotive | High Conductance Thermal Straps for Cryocooler & Cryogenics", Thermotive.com, 2022. [Online]. Available: http://www.thermotive.com/thermalstraps.html. [Accessed: 03- Jun- 2022].

[21]   Larid, "Tflex HD80000 Series Thermal Gap Filler". 2011.

[22]   A. Lécossais, F.Jacquemart, G. Lefort, E. Dehombreux, Deployable Panel Radiator. 2015.

[23]   S. Selvadurai, A. Chandran, D. Valentini and B. Lamprecht, "Passive Thermal Control Design Methods, Analysis, Comparison, and Evaluation for Micro and Nanosatellites Carrying Infrared Imager", Applied Sciences, vol. 12, no. 6, p. 2858, 2022. Available: 10.3390/app12062858.

[24]   D. Gilmore, Spacecraft thermal control handbook. Norwich, NY: Knovel, 2007

[25]   Thermal Management of High Power Applications in CubeSat. [Online]. Available: http://dx.doi.org/10.13140/RG.2.2.15695.94881. [Accessed: 03- Jun- 2022].

[26]   K.F Sam, C.H, Z. Deng, Optimization of a space based radiator. 2011.

[27]   K. Arif, C. Roberto, Small Spacecraft State of the Art Report. 2015.

[28]   D. Gilmore, Spacecraft thermal control handbook. Norwich, NY: Knovel, 2007

[29]   K. R.D, Satellite Thermal Control for Systems Engineers. 1998.

[30]   H. Brower, J. Guo, Solving the Thermal Challenge in Power-Dense CubeSats with

water Heat Pipes. 2006.

[31]   I. Garmendia, H. Vallejo, Design and Fabrication of a Phase Change Material Heat Storage Device for the Thermal Control of Electronics Components of Space Applications. 2022.

[32]   R. Hon, C. Kesler, D. Sigurdson, Integrated Testing of a Complete Low Cost Space Cryocooler System. 2009.

[33]   "7.0 Thermal Control", NASA, 2022. [Online]. Available: https://www.nasa.gov/smallsat-institute/sst-soa/thermal-control. [Accessed: 03- Jun- 2022].

[34]   J. Olson, CubeSat-Sized Space Microcryocooler. 2020.

[35]   Sunpower Inc, CryoTel® DS 1.5. 2015.

[36]   D. Durand, E. Tward, G. Toma, T. Nguyen. Efficient High Capacity Space Microcooler. 2014.

[37]   "LPT95101400 mW / 18 mm slip-on - Thales Cryogenics", Thales Cryogenics, 2022. [Online]. Available: https://www.thales-cryogenics.com/products/coolers/lpt/lpt9510/. [Accessed: 03- Jun- 2022].

[38]   M. Farison, K. Hicks, M. Schmidt, S. YangCloud, CubeSat Thermoelectric Cooler Controller 1 System. 2010.

[39]   "7.0 Thermal Control", NASA, 2022. [Online]. Available: https://www.nasa.gov/smallsat-institute/sst-soa/thermal-control. [Accessed: 03- Jun- 2022].

[40]   G.J Snyder, E.S.Toberer, Comply thermoelectric materials. 2008.

[41]   M. Le Berre, , S. Launay, Fabrication and experimental investigation of silicon micro heat pipes for cooling electronics.

[42]   C. Samo, J.B. Dezord, Use of Metal Matrix Composite Material Heat Pipes For The Thermal Management of High Integrated Electronic Packages. 1999.

[43]   S.W. Kang, S.H. Tsai, M.H. Ko, Metallic micro heat pipe heat spreader fabrication, Applied Thermal Engineering. 2004.

[44]   M. Marengo, S. Zhdanov, Micro-Heat-Sinks for Space Applications. 2004.

[45]   P. Champagne, Development of a J-T Micro Compressor. 2015.

[46]   T. Ganzeboom, H.S.B. Brower, Mini Mechanically Pumped Loop Modelling and Design for standardized CubeSat thermal control. 2020.